ERNST CHLADNI'S SOUND EXPERIMENTS
CLANG!
Written by
Darcy Pattison
Illustrated by
Peter Willis
N

Clang! Ernst Chladni's
Sound Experiments
by Darcy Pattison
Illustrated by Peter Willis

Mims House
1309 Broadway
Little Rock, AR 72202

MimsHouse.com

Publisher's Cataloging-in-Publication data

Names: Pattison, Darcy, author. | Willis, Peter N., illustrator.
Titles: CLANG! Ernst Chladni's sound experiments / written by Darcy Pattison; illustrated by Peter Willis.

Description: Little Rock, AR: Mims House, 2018.
Identifiers: ISBN 978-1-62944-093-4 (Hardcover) | 978-1-62944-094-1 (pbk.) | 978-1-62944-095-8 (ebook) | LCCN 2017915973
Subjects: LCSH Chladni, Ernst Florens Friedrich, 1756-1827--Juvenile literature. | Sound--Juvenile literature. | Physicists--Germany--Biography. | Music--Acoustics and physics--19th century--Juvenile literature. | Germany--History--19th century. | BISAC JUVENILE NONFICTION / Biography & Autobiography / Historical | JUVENILE NONFICTION / Biography & Autobiography / Science & Technology
Classification: LCC Q[illegible] P38 [illegible] | DDC [illegible]092--dc23

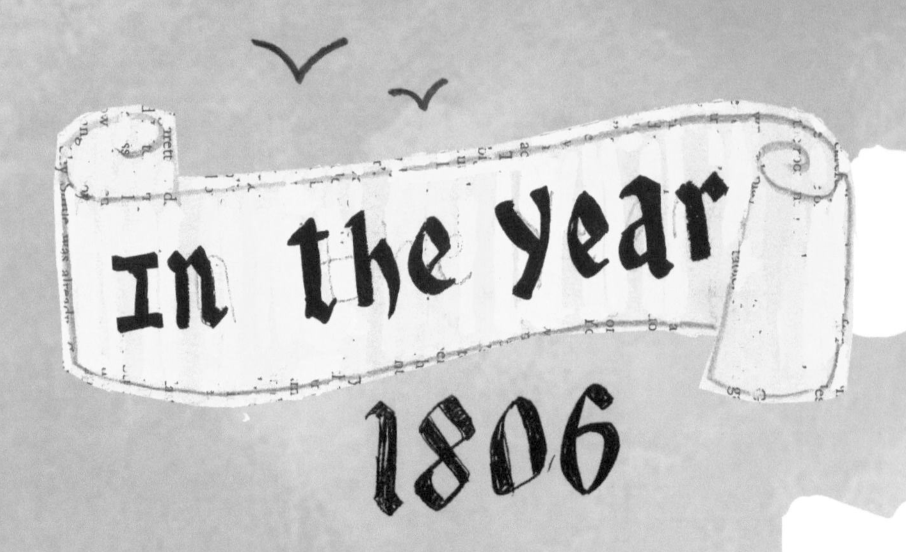

Ernst Florens Frieddrich Chladni (KLOD-Nee) locked the door of his house in Wittenberg, Germany, . . .

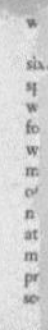

DR
CHLADNI

. . . climbed into his spacious carriage, and left town. It would be four years before he returned. Chladni was a scientist who studied sound. Unlike many scientists of the time, Chladni didn't work for a university. Instead, he traveled because he earned money by entertaining people with his sound experiments.

Chladni traveled in the Netherlands and Brussels for two years. In 1808, Chladni moved on to Paris, France. There he met with French scientists who admired his work.

The French scientists especially liked his 1802 book, *Die Akustik*, or *The Acoustics.* The book explained the science of sound and sound waves. However, the book was written in Chladni's native German language. French scientists encouraged Chladni to translate the book into French.

Die
Akustik

But Chladni had two problems. First, he needed enough money to live on while he did his work.

Second, he needed someone to make sure his French was correct.

Chladni's friends, the French scientists, decided to help with both problems.

At 7 pm on a Tuesday evening in February 1809, a carriage pulled up to the Tuilerie Palace in Paris, France. This was the home of the notorious Napoleon Bonaparte, Emperor of the French people. Perhaps Emperor Bonaparte would help fund Chladni's book.

When Chladni entered, Napoleon welcomed him. He introduced Chladni to his wife, his mother, his uncle and his assistant.

Quickly, Chladni set up his equipment. First, he entertained them by playing the clavicylinder, a musical instrument he had invented. It looked something like a piano, but pianos create sound by hitting wires of different lengths.

Instead, the clavicylinder made sound by rotating a glass cylinder. When he pushed a key on a keyboard, a bar moved forward to touch the cylinder. The bar vibrated and made a sound. Chladni played some musical pieces on the clavicylinder. It pleased everyone.

Napoleon asked if he could play the clavicylinder. Chladni agreed, but he suggested that it should be played softly because of the glass cylinder. Instead, Napoleon banged the keyboard! Fortunately, the clavicylinder didn't break.

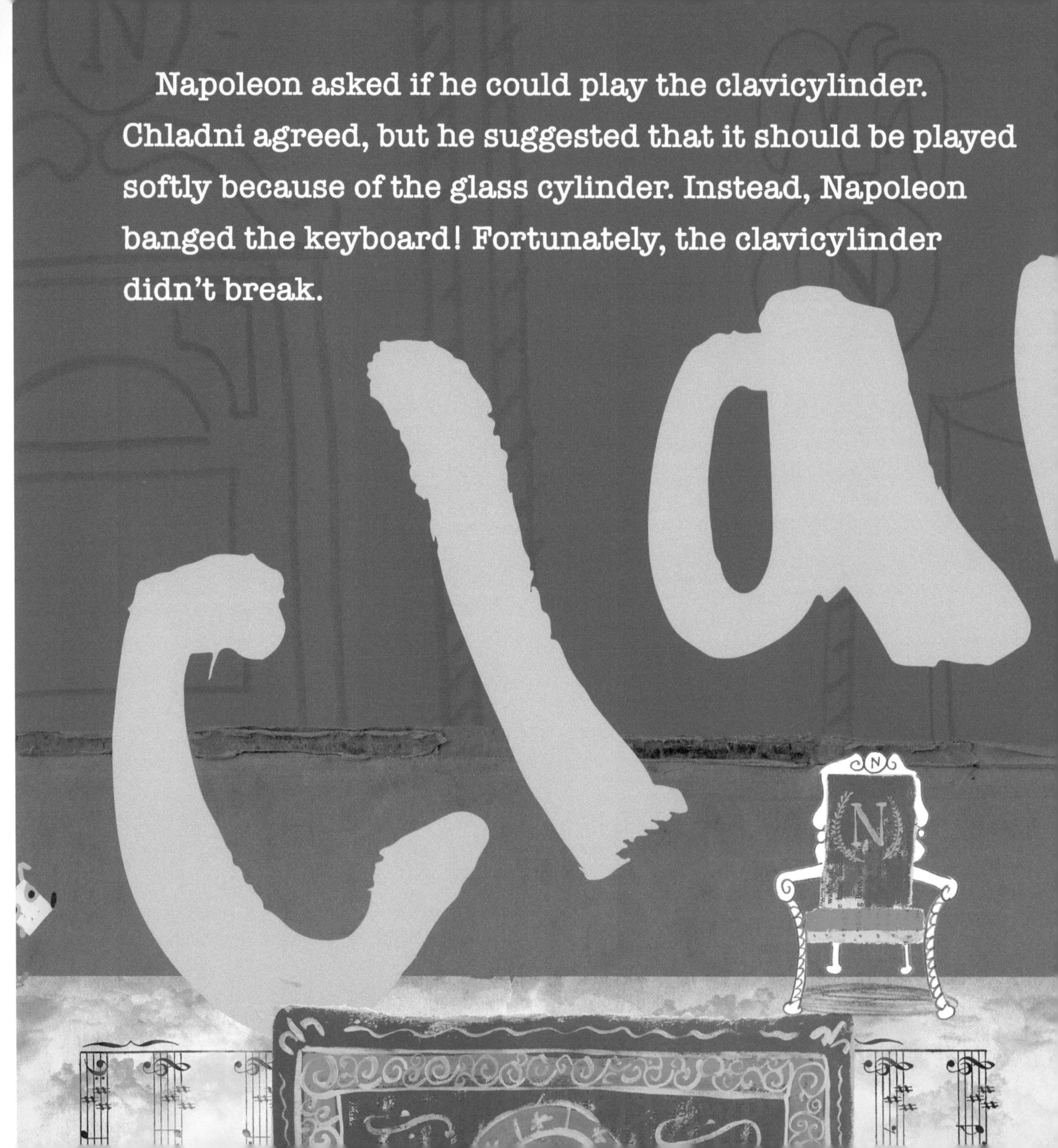

clang!
clang!
clang!
clang!

Napoleon whispered to one of the French scientists:
Could I see the inside of Clavicylinder?

Chladni overheard the Emperor's question. He answered:

He wouldn't allow anyone to look inside because he was afraid someone might steal his invention.

Next, Chladni planned to demonstrate sound experiments. Sound waves are hard to study because they can't be seen. Early in his career, Chladni studied how sound was created by vibrating strings of a musical instrument. Next, he studied how fast sound waves moved through the air of a pipe organ. His most famous experiments, though, were how sound traveled through solids, such as metals or glass.

Guitar - Vibrating Strings

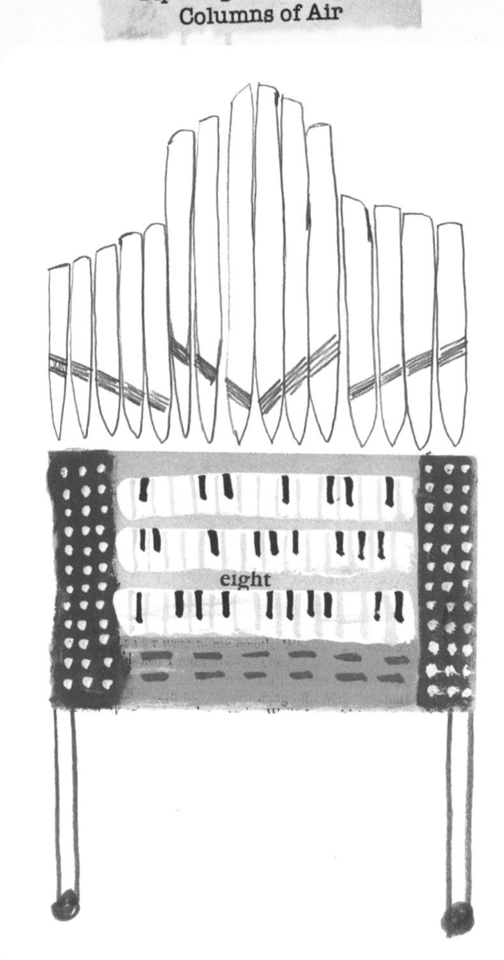

Pipe Organ - Vibrating Columns of Air

Chladni started his sound experiment for the Emperor by centering a brass plate onto a single support. Sand was sprinkled over the plate's surface. He used a violin bow to stroke the side of the brass plate. As the sound wave traveled through the plate, it made the brass vibrate. The vibrations pushed the sand around. Sand gathered in places where the brass plate was not vibrating. Very low sounds made simple shapes. Very high sounds made complex shapes.

Metal Plate - Vibrating
Solid

The audience was fascinated by the sound wave figures in the sand.

Napoleon knew that mathematics can explain science. He asked Chladni to explain the mathematics of how sound waves travel through solids. While Chladni understood some of it, he didn't have the full mathematical explanation.

Chladni's Sound Wave Figures Made by Sand

TRAITE
d'acoustique
1809
by
Chladni
6000
FRENCH FRAN

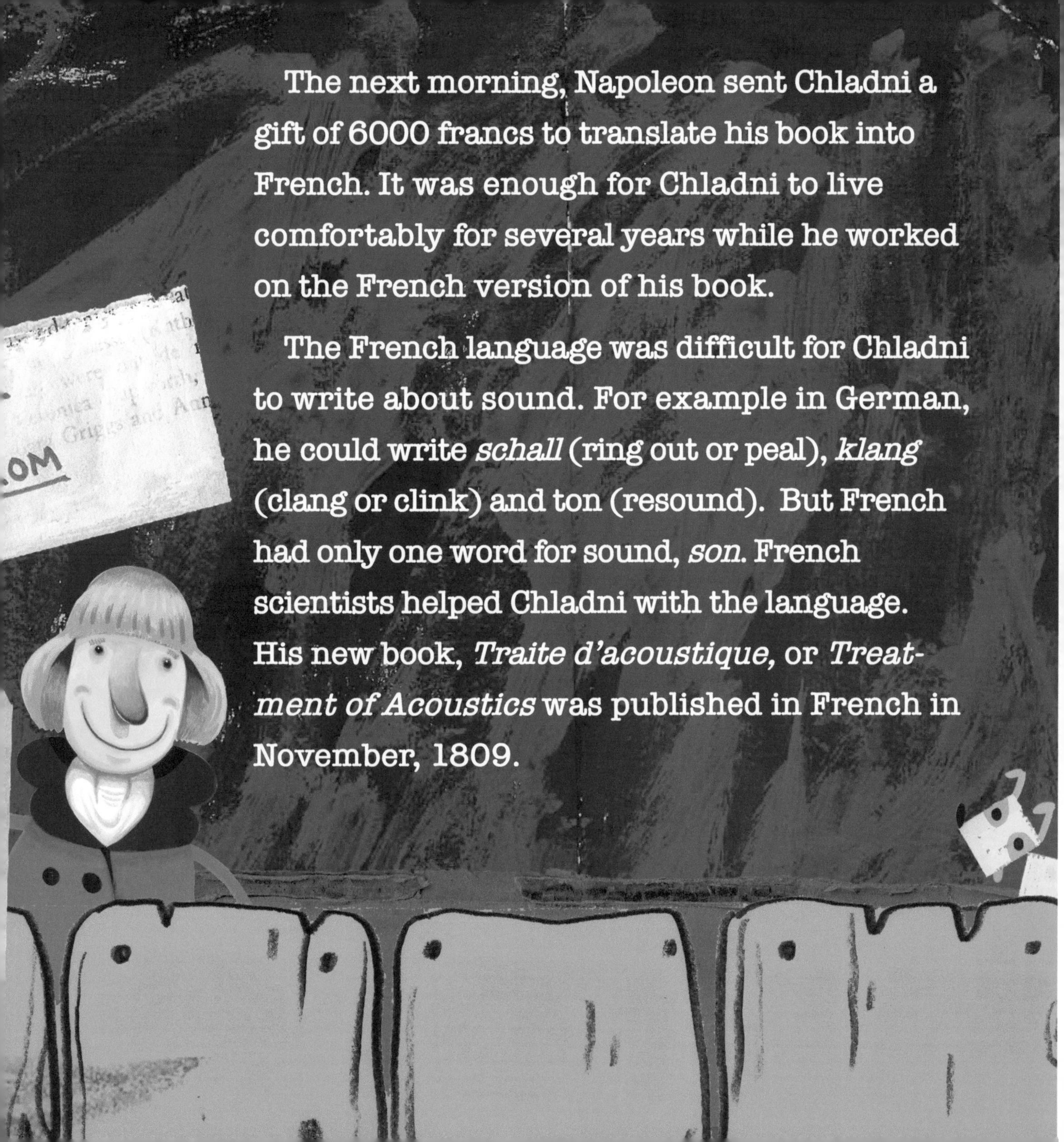

The next morning, Napoleon sent Chladni a gift of 6000 francs to translate his book into French. It was enough for Chladni to live comfortably for several years while he worked on the French version of his book.

The French language was difficult for Chladni to write about sound. For example in German, he could write *schall* (ring out or peal), *klang* (clang or clink) and ton (resound). But French had only one word for sound, *son*. French scientists helped Chladni with the language. His new book, *Traite d'acoustique,* or *Treatment of Acoustics* was published in French in November, 1809.

In the opening, Chladni wrote this dedication:

Napoleon the Great deigned to accept the dedication of this work after having seen its fundamental experiments.

Traite
d'acousti
180
DR. CHL

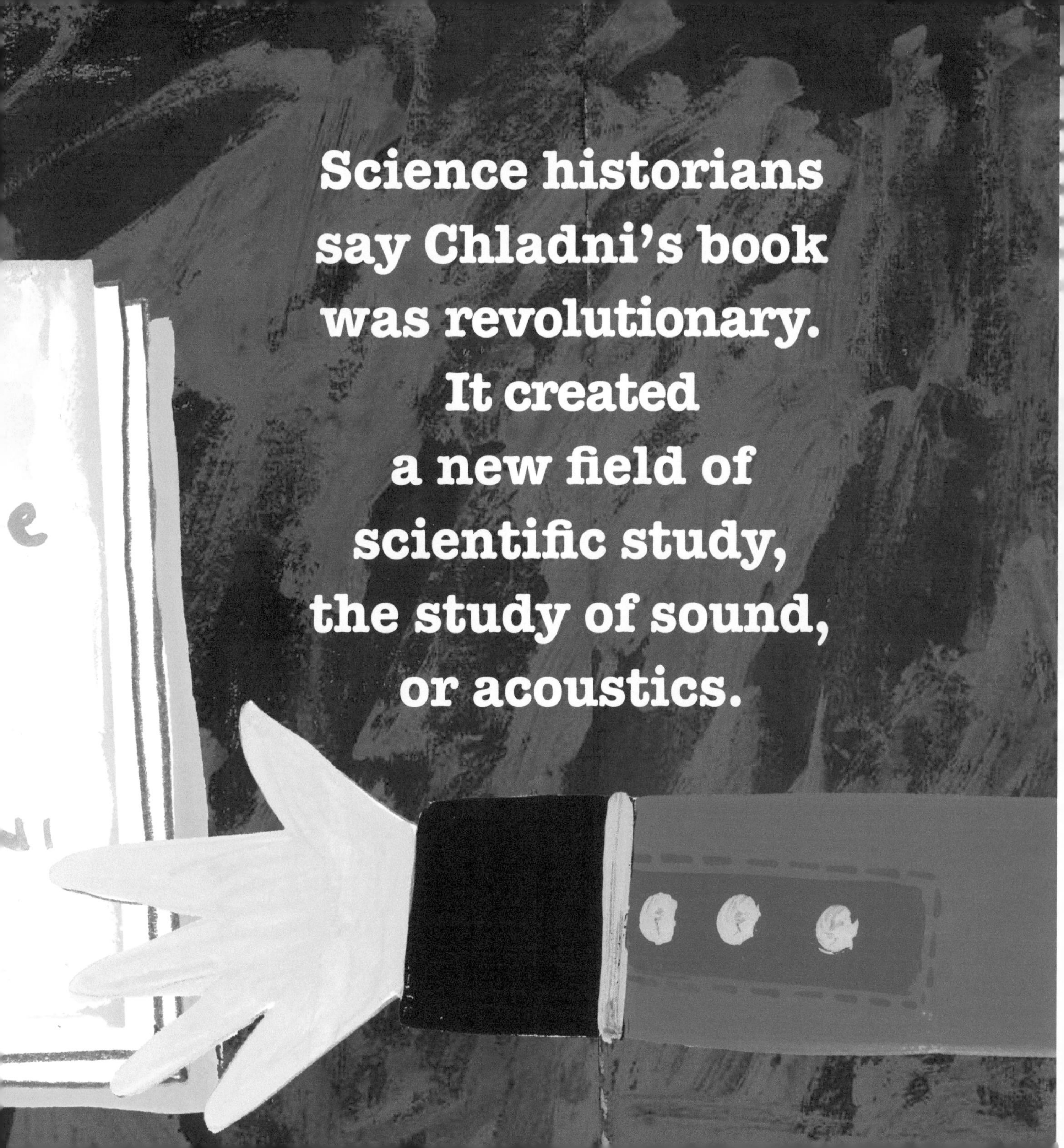

Science historians
say Chladni's book
was revolutionary.
It created
a new field of
scientific study,
the study of sound,
or acoustics.

son!
son!
son!
son!
TRAITE
d'acoustique
1809
by
Chladni

WHAT IS SOUND?

Let's say that you stretch a rubber band across a box. When you pluck the rubber band, you'll hear a sound. Notice that the rubber band does not make the sound. Instead, it's made when the rubber band vibrates because you plucked it. The vibrations create waves in the air. The human ear picks up the waves and interprets them as sound.

When Chladni studied sound, he studied how vibrations could be made by making different things vibrate. He studied vibrating columns of air by studying pipe organs. He studied vibrating wires by looking at musical instruments such as a guitar or violin. Finally, he studied vibrating solids by studying vibrations of metal plates or glass.

ERNST CHLADNI, THE FATHER OF ACOUSTICS

Ernst Florens Friedrich Chladni (KLOD-nee) (1756 – 1827) was a German scientist and entertainer. He used scientific experiments and music to entertain and teach people across Europe. His sound experiments brought him fame. In 1802, he published *Die Akustik* (*The Acoustics*) in his native language, German. Friends and fellow scientists in Paris asked him to translate the book into French.

Chladni wasn't like many scientists of the time, though. He didn't work for a university, which meant he needed money from other sources. In February, 1806, scientists introduced Chladni to Emperor Napoleon. The Emperor's support meant he could spend the time needed to finish his French book, *Traite d'acoustique* (*Treatment of Acoustics*), which earned him the title, the "Father of Acoustics." Scientists have always cooperated across international boundaries, and often progress only happens when this type of cooperation occurs.

THE CLAVICYLINDER – Chladni's Musical Instrument

The clavicylinder made music by rotating a glass cylinder and then moving tuned rods to touch the cylinder and set off the sound. Because the central cylinder was glass, it was a delicate instrument to play. Only a few instruments are still available in museums.

THE FATHER OF METEORITICS

As Chladni traveled across Europe, he also collected over 40 meteorites. With this widespread collection, he could make new and interesting observations. He was the first to propose that meteorites came from space. In his later years, he included the meteorites in his lectures and demonstrations. Some people call Chaldni the "Father of Meteoritics." A small lunar impact crater is named Chladni in his honor.

THE MATH OF SOUND

After talking with Chladni, Emperor Napoleon also set up a prize of 3000 francs for whomever could write a mathematical theory of sound waves. The prize was won in 1816 by French mathematician Sophie Germain, the first woman to win a prize from the Paris Academy of Sciences.

SOURCES

After the audience with Napoleon, Chladni wrote an article about the evening with Napoleon for the *Caecilia*, a German music magazine. The article was quoted in his 1888 biography by Franz Melde, *Chladni: Life and Times*. This story is based on Chladni's own account of his meeting with Napoleon.

CPSIA information can be obtained
at www.ICGtesting.com
Printed in the USA
LVHW07n1744310518
579127LV00015B/303/P